Additional Topics in Anima
Graphics, and Simulink

A Supplement to

Introduction to MATLAB® 6 for Engineers

William J. Palm III
University of Rhode Island

Higher Education

Boston   Burr Ridge, IL   Dubuque, IA   Madison, WI   New York
San Francisco   St. Louis   Bangkok   Bogotá   Caracas   Kuala Lumpur
Lisbon   London   Madrid   Mexico City   Milan   Montreal   New Delhi
Santiago   Seoul   Singapore   Sydney   Taipei   Toronto

The **McGraw·Hill** Companies

Additional Topics in Animations, Graphics, and Simulink®
A supplement to INTRODUCTION TO MATLAB® 6 FOR ENGINEERS
WILLIAM J. PALM III

Published by McGraw-Hill Higher Education, an imprint of The McGraw-Hill Companies, Inc.,
1221 Avenue of the Americas, New York, NY 10020. Copyright © 2001 by The McGraw-Hill
Companies, Inc. All rights reserved.

This book will be printed on acid-free paper.

1 2 3 4 5 6 7 8 9 0 CUS CUS 0 9 8 7 6 5 4 3

ISBN 0-07-296791-9

www.mhhe.com

Additional Topics in Animation, Graphics, and Simulink®

A Supplement to

Introduction to MATLAB®6 for Engineers

William J. Palm III

University of Rhode Island

Table of Contents

This is a second supplement to *Introduction to MATLAB 6 for Engineers*, (Ref. 1), herein referred to as "the textbook." This supplement contains topics, not covered in the textbook, that have seen increased interest among users since the publication of the textbook in 2001. Another supplement, titled *MATLAB 6.5 Supplement*, was published in 2002 to coincide with the release of MATLAB 6.5. That supplement is intended to update the user on the new features in MATLAB 6.5 that are relevant to the topics contained in the textbook, and it expands on certain topics whose importance has increased since the publication of the textbook.

In the following sections we reference the relevant chapters in the textbook (Ref. 1).

1. Creating Movies in MATLAB

[Reference: Chapter 5]

Animation can be used to display the behavior of an object over time. Some of the MATLAB demos are M-files that perform animation. After completing this section and the next, which have simple examples, you may study the demo files, which are more advanced. Two methods can be used to create animations in MATLAB. The first method uses the `movie` function and is discussed in this section. The second method uses the EraseMode property, which is discussed in Section 2.

1.1 The `getframe` and `movie` Functions

The `getframe` command captures, or takes a snapshot of, the current figure to create a single frame for the movie. The `getframe` function is usually used in a `for` loop to assemble an array of movie frames. The `movie` function plays back the frames after they have been captured.

To create a movie, use a script file of the following form.

```
for k = 1:n
    plotting expressions
    M(k) = getframe;    % Saves current figure in array M
end
movie(M)
```

For example, the following script file creates 20 frames of the function $te^{-t/b}$ for $0 \leq t \leq 100$ for each of 20 values of the parameter b from $b = 1$ to $b = 20$.

```
% Program movie1.m
% Animates the function t*exp(-t/b).
t = [0:0.05:100];
for b = 1:20
    plot(t,t.*exp(-t/b)),axis([0 100 0 10]),xlabel('t');
    M(:,b) = getframe;
end
```

The line `M(:,b) = getframe;` acquires and saves the current figure as a column of the matrix M. Once this file is run, the frames can be replayed as a movie by typing `movie(M)`. The animation shows how the location and height of the function peak changes as the parameter b is increased.

1.2 Rotating a 3-D Surface

The following example rotates a three-dimensional surface by changing the viewpoint. The data is created using the built-in function, **peaks**.

2

```
% Program movie2.m
% Rotates a 3-D surface.
[X,Y,Z] = peaks(50);        % Create data.
surfl(X,Y,Z)        % Plot the surface.
axis([-3 3 -3 3 -5 5]) % Retain same scaling for each frame.
axis vis3d off % Set the axes to 3D and turn off tick marks,
                % and so forth.
shading interp      % Use interpolated shading.
colormap(winter)    % Specify a colormap.
for k = 1:30 % Rotate the viewpoint and capture each frame.
   view(-37.5+0.5*(k-1),30)
   M(k) = getframe;
end
cla      % Clear the axes.
movie(M)     % Play the movie.
```

The colormap(map) function sets the current figure's color map to map. Type help graph3d to see a number of colormaps to choose for map. The choice winter provides blue and green shading. The view function specifies the 3-D graph viewpoint. The syntax view(az,el) sets the angle of the view from which an observer sees the current 3-D plot, where az is the azimuth or horizontal rotation and el is the vertical elevation (both in degrees). Azimuth revolves about the z-axis, with positive values indicating counter-clockwise rotation of the viewpoint. Positive values of elevation correspond to moving above the object; negative values move below. The choice az = -37.5, el = 30 is the default 3-D view.

1.3 Extended Syntax of the movie Function

The function movie(M) plays the movie in array M once, where M must be an array of movie frames (usually acquired with getframe). The function movie(M,n) plays the movie n times. If n is negative, each "play" is once forward and once backward. If n is a vector, the first element is the number of times to play the movie, and the remaining elements are a list of frames to play in the movie. For example, if M has four frames, then n = [10 4 4 2 1] plays the movie ten times, and the movie consists of frame 4 followed by frame 4 again, followed by frame 2 and, finally, frame 1.

The function movie(M,n,fps) plays the movie at fps frames per second. If fps is omitted, the default is 12 frames per second. Computers that cannot

3

achieve the specified `fps` will play the movie as fast as they can. The function `movie(h,...)` plays the movie in object `h`, where `h` is a handle to a figure or an axis. Handles are discussed in Section 2.2.

The function `movie(h,M,n,fps,loc)` specifies the location to play the movie, relative to the lower-left corner of object `h` and in pixels, regardless of the value of the object's Units property, where `loc = [x y unused unused]` is a four-element position vector, of which only the x and y coordinates are used, but all four elements are required. The movie plays back using the width and height in which it was recorded.

Note that for code to be compatible with versions of MATLAB prior to Release 11 (5.3), the `moviein(n)` function must be used to initialize movie frame memory for n frames. To do this, place the line `M = moviein(n);` before the `for` loop that generates the plots.

The disadvantage of the `movie` function is that it might require too much memory if many frames or complex images are stored.

2. Animation with the EraseMode Property

[Reference: Chapter 5]

One form of extended syntax for the `plot` function is

  `plot(...,'PropertyName','PropertyValue',...)`

This form sets the plot property specified by `PropertyName` to the values specified by `PropertyValue` for all line objects created by the `plot` function. One such property name is EraseMode. This property controls the technique MATLAB uses to draw and erase line objects and is useful for creating animated sequences. The allowable values for the EraseMode property are the following.

- `normal` This is the default value for the EraseMode property. By typing

  `plot(...,'EraseMode','normal'))`

  the entire figure, including axes, labels, and titles, is erased and redrawn using only the new set of points. In redrawing the display, a three-dimensional analysis is performed to ensure that all objects are rendered correctly. Thus, this mode produces the most accurate picture but is the slowest. The other modes are faster but do not perform

4

a complete redraw and are therefore less accurate. This method may cause blinking between each frame because everything is erased and redrawn. This method is therefore undesirable for animation.

- **none** When the EraseMode property value is set to **none**, objects in the existing figure are not erased, and the new plot is superimposed on the existing figure. This mode is therefore fast because it does not remove existing points, and it is useful for creating a "trail" on the screen.

- **xor** When the EraseMode property value is set to **xor**, objects are drawn and erased by performing an exclusive OR with the background color. This produces a smooth animation. This mode does not destroy other graphics objects beneath the ones being erased and does not change the color of the objects beneath. However, the object's color depends on the background color.

- **background** When the EraseMode property value is set to **background**, the result is the same as with **xor** except that objects behind the erased objects are destroyed. Objects are erased by drawing them in the axes' background color or in the figure background color if the axes Color property is set to **none**. This damages objects that are behind the erased line, but lines are always properly colored.

The **drawnow** command causes the previous graphics command to be executed immediately. If the **drawnow** command were not used, MATLAB would complete all other operations before performing any graphics operations and would display only the last frame of the animation.

Animation speed depends of the intrinsic speed of the computer and on what and how much is being plotted. Symbols such as o, *, or + will be plotted slower than a line. The number of points being plotted also affects the animation speed. The animation can be slowed by using the **pause(n)** function, which pauses the program execution for n seconds.

2.1 Using Object Handles

An expression of the form

```
p = plot(...)
```

assigns the results of the plot function to the variable p, which is a figure identifier called a "figure handle." This stores the figure and makes it available for future use. Any valid variable name may be assigned to a handle. A figure handle is a specific type of "object handle." Handles may be assigned to other types of objects. For example, later we will create a handle with the text function.

The set function can be used with the handle to change the object properties. This function has the general format

```
set(object handle, 'PropertyName', 'PropertyValue', ...)
```

If the object is an entire figure, its handle also contains the specifications for line color and type, marker size, and the value of the EraseMode property. Two of the properties of the figure specify the data to be plotted. Their property names are XData and YData. The following example shows how to use these properties.

2.2 Animating a Function

Consider the function $te^{-t/b}$, which was used in the first movie example. This function can be animated as the parameter b changes with the following program.

```
% Program animate1.m
% Animates the function t*exp(-t/b).
t = [0:0.05:100];
b = 1;
p = plot(t,t.*exp(-t/b),'EraseMode','xor');...
    axis([0 100 0 10]),xlabel('t');
for b = 2:20
    set(p,'XData',t,'YData',t.*exp(-t/b)),...
    axis([0 100 0 10]),xlabel('t');
    drawnow
    pause(0.1)
end
```

In this program the function $te^{-t/b}$ is first evaluated and plotted over the range $0 \le t \le 100$ for $b = 1$, and the figure handle is assigned to the variable p. This establishes the plot format for all following operations, for example, line type and color, labeling, and axis scaling. The function $te^{-t/b}$ is then evaluated and plotted over the range $0 \le t \le 100$ for $b = 2$, 3, 4, ... in the `for` loop, and the previous plot is erased. Each call to `set` in the `for` loop causes the next set of points to be plotted. The EraseMode property value specifies how to plot the existing points on the figure (i.e., how to refresh the screen), as each new set of points is added. You should investigate what happens if the EraseMode property value is set to `none` instead of `xor`.

2.3 Animating Projectile Motion

This following program illustrates how user-defined functions and subplots can be used in animations. The following are the equations of motion for a projectile launched with a speed s_0 at an angle θ above the horizontal, where x and y are the horizontal and vertical coordinates, g is the acceleration due to gravity, and t is time.

$$x(t) = (s_0 \cos \theta) t \qquad\qquad y(t) = -\frac{gt^2}{2} + (s_0 \sin \theta) t$$

By setting $y = 0$ in the second expression, we can solve for t and obtain the following expression for the maximum time the projectile is in flight, t_{max}.

$$t_{max} = \frac{2s_0}{g} \sin \theta$$

The expression for $y(t)$ may be differentiated to obtain the expression for the vertical velocity:

$$v_{vert} = \frac{dy}{dt} = -gt + s_0 \sin \theta$$

The maximum distance x_{max} may be computed from $x(t_{max})$, the maximum height y_{max} may be computed from $y(t_{max}/2)$, and the maximum vertical velocity occurs at $t = 0$.

The following functions are based on these expressions, where $s0$ is the launch speed s_0, and `th` is the launch angle θ.

```
function x = xcoord(t,s0,th);
% Computes projectile horizontal coordinate.
x = s0*cos(th)*t;
```

```
function y = ycoord(t,s0,th,g);
% Computes projectile vertical coordinate.
y = -g*t.^2/2+s0*sin(th)*t;



function v = vertvel(t,s0,th,g);
% Computes projectile vertical velocity.
v = -g*t+s0*sin(th);
```

The following program uses these functions to animate the projectile motion in the first subplot, while simultaneously displaying the vertical velocity in the second subplot, for the values $\theta = 45°$, $s_0 = 105$ feet/second, and $g = 32.2$. Note that the values of xmax, ymax, and vmax are computed and used to set the axes scales. The figure handles are h1 and h2.

```
% Program animate2.m
% Animates projectile motion.
% Uses functions xcoord, ycoord, and vertvel
th = 45*(pi/180);
g = 32.2; s0 = 105;
%
tmax = 2*s0*sin(th)/g;
xmax = xcoord(tmax,s0,th);
ymax = ycoord(tmax/2,s0,th,g);
vmax = vertvel(0,s0,th,g);
w = linspace(0,tmax,500);
%
subplot(2,1,1)
plot(xcoord(w,s0,th),ycoord(w,s0,th,g)),hold,
h1 = plot(xcoord(w,s0,th),ycoord(w,s0,th,g),'o','EraseMode','xor');
axis([0 xmax 0 1.1*ymax]),xlabel('x'),ylabel('y')
subplot(2,1,2)
plot(xcoord(w,s0,th),vertvel(w,s0,th,g)),hold,
h2 = plot(xcoord(w,s0,th),vertvel(w,s0,th,g),'s','EraseMode','xor');
   axis([0 xmax 0 1.1*vmax]),xlabel('x'),...
   ylabel('Vertical Velocity')
for t = [0:0.01:tmax]
   set(h1,'XData',xcoord(t,s0,th),'YData',ycoord(t,s0,th,g))
   set(h2,'XData',xcoord(t,s0,th),'YData',vertvel(t,s0,th,g))
```

```
        drawnow
        pause(0.005)
    end
    hold
```

You should experiment with different values of the **pause** function argument.

2.4 Animation with Arrays

Thus far we have seen how the function to be animated may be evaluated in the **set** function with an expression or with a function. A third method is to compute the points to be plotted ahead of time and store them in arrays. The following program shows how this is done, using the projectile application. The plotted points are stored in the arrays x and y.

```
% Program animate3.m
% Animation of a projectile using arrays.
th = 70*(pi/180);
g = 32.2; s0=100;
tmax = 2*s0*sin(th)/g;
xmax = xcoord(tmax,s0,th);
ymax = ycoord(tmax/2,s0,th,g);
%
w = linspace(0,tmax,500);
x = xcoord(w,s0,th);y = ycoord(w,s0,th,g);
plot(x,y),hold,
h1 = plot(x,y,'o','EraseMode','xor');
axis([0 xmax 0 1.1*ymax]),xlabel('x'),ylabel('y')
%
kmax = length(w);
for k =1:kmax
    set(h1,'XData',x(k),'YData',y(k))
    drawnow
    pause(0.001)
end
hold
```

2.5 Displaying Elapsed Time

It may be helpful to display the elapsed time during an animation. To do this, modify the program animate3.m as shown in the following. The new lines are indicated in bold; the line formerly below the line h1 = plot(... has been deleted.

```
% Program animate4.m
% Like animate3.m but displays elapsed time.
th = 70*(pi/180);
g = 32.2; s0 = 100;
%
tmax = 2*s0*sin(th)/g;
xmax = xcoord(tmax,s0,th);
ymax = ycoord(tmax/2,s0,th,g);
%
t = linspace(0,tmax,500);
x = xcoord(t,s0,th);y = ycoord(t,s0,th,g);
plot(x,y),hold,
h1 = plot(x,y,'o','EraseMode','xor');
text(10,10,'Time = ')
time = text(30,10,'0','EraseMode','background')
    axis([0 xmax 0 1.1*ymax]),xlabel('x'),ylabel('y')
%
kmax = length(t);
for k = 1:kmax
    set(h1,'XData',x(k),'YData',y(k))
    t_string = num2str(t(k));
    set(time,'String',t_string)
    drawnow
    pause(0.001)
end
hold
```

The first new line creates a label for the time display using the text statement, which writes the label once. The program must not write to that location again. The second new statement creates the handle time for the text label and creates the string for the first time value, which is 0. By using the background value for EraseMode, the statement specifies that the existing display of the time variable will be erased when the next value is

10

displayed. Note that the numerical value of time `t(k)` must be converted to a string, by using the function `num2str`, before it can be displayed. In the last new line, in which the `set` function uses the `time` handle, the property name is `'String'`, which is not a variable but a property associated with text objects. The variable being updated is `t_string`.

3. Additional Simulink Examples

[Reference: Chapter 8]

The popularity of Simulink has greatly increased since the publication of the textbook in 2001, as evidenced by the numerous short courses offered at meetings sponsored by professional organizations such as the ASEE. Section 8.9 of the textbook is an introduction to Simulink, and it covers some of the basic building blocks in Simulink. We assume that you are familiar with the topics covered in that section. Here we introduce some additional blocks that are often useful for simulation.

The Simulink blocks are located in "libraries," as discussed in the textbook. As Simulink evolves through new versions, some libraries are renamed and some blocks are moved to different libraries, so we will not specify the library here. The best way to locate a block, given its name, is to type its name in the Find pane at the top of the Simulink Library Browser. When you press Enter, Simulink will take you to the block location and will display a brief description of the block in the pane below the Find pane.

We discuss the following Simulink features in this section.

- Subsystem blocks,

- Input and Output Ports,

- the Constant block,

- the Transport Delay block,

- specifying initial conditions with a Transfer Function block,

- the Saturation block,

- the Rate Limiter block,

- the Derivative block,

- the Signal Builder block,

- the Look-Up Table block, and

- the MATLAB Fcn block.

Most of these blocks have several property values, and we will discuss only those values that are pertinent to the example. Unless otherwise stated, in the following examples you should use the default values for the block properties.

We will use two physical applications to illustrate these features. Their models are based in simple physical principles that are familiar to engineers. The first is a hydraulic system, whose model is based on conservation of mass, and the second is a vehicle suspension system, whose model is based on Newton's second law. Because their governing equations are similar to other engineering applications, such as electric circuits and devices, the lessons learned from these examples will enable you to use Simulink for other applications.

3.1 A Hydraulic Example

To provide a physical example whose behavior is easy to visualize, we will study a system composed of a tank of liquid of mass density ρ (Figure 3.1). The tank shown in cross section in the figure is cylindrical with a bottom area A. A flow source dumps liquid into the tank at the mass flow rate $q_i(t)$. The total mass in the tank is $m = \rho A h$, and from conservation of mass we have

$$\frac{dm}{dt} = \rho A \frac{dh}{dt} = q_i - q_o \qquad (1)$$

since ρ and A are constants.

If the outlet is a pipe that discharges to atmospheric pressure p_a and provides a resistance to flow that is proportional to the pressure difference across its ends, then the outlet flow rate is

$$q_o = \frac{1}{R}\left[(\rho g h + p_a) - p_a\right] = \frac{\rho g h}{R}$$

where R is called the *fluid resistance*. Substituting this expression into equation (1) gives the model:

$$\rho A \frac{dh}{dt} = q_i(t) - \frac{\rho g}{R}h$$

12

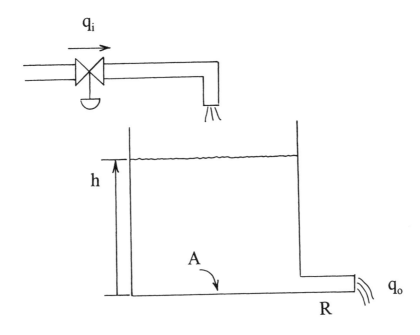

Figure 3.1: A hydraulic system with flow input.

The transfer function is

$$\frac{H(s)}{Q_i(s)} = \frac{1}{\rho A s + \rho g / R} = \frac{R/\rho g}{\frac{RA}{g} s + 1}$$

On the other hand, the outlet may be a valve or other restriction that provides nonlinear resistance to the flow. In such cases, a common model is the signed-square-root relation

$$q_o = \frac{1}{R} \text{SSR}(\Delta p)$$

where q_o is the outlet mass flow rate, R is the resistance, Δp is the pressure difference across the resistance, and

$$\text{SSR}(\Delta p) = \begin{cases} \sqrt{\Delta p} & \text{if } \Delta p \geq 0 \\ -\sqrt{|\Delta p|} & \text{if } \Delta p < 0 \end{cases}$$

Note that we may express the SSR(u) function in MATLAB as follows: `sgn(u)*sqrt(abs(u))`.

13

Consider the slightly different system shown in Figure 3.2, which has a flow source q and two pumps that supply liquid at the pressures p_l and p_r. Suppose the resistances are nonlinear and obey the signed-square-root relation. Then the model of the system is the following:

$$\rho A \frac{dh}{dt} = q + \frac{1}{R_l} \text{SSR}(p_l - p) - \frac{1}{R_r} \text{SSR}(p - p_r)$$

where A is the bottom area and $p = \rho g h$. The pressures p_l and p_r are the *gage* pressures at the left- and right-hand sides. Gage pressure is the difference between the absolute pressure and atmospheric pressure. Note that the atmospheric pressure p_a cancels out of the model because of the use of gage pressure.

We will use this application to introduce the following Simulink elements:

- Subsystem blocks,

- Input and Output Ports, and

- the Constant block.

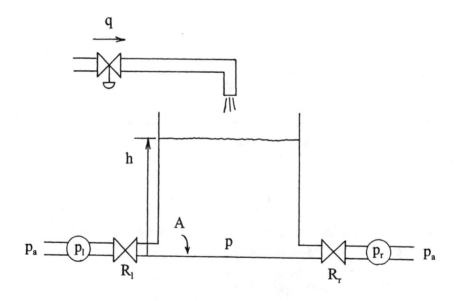

Figure 3.2: A hydraulic system with flow and pressure inputs.

3.2 Subsystem Blocks

One potential disadvantage of a graphical interface such as Simulink is that to simulate a complex system, the diagram can become rather large and therefore, somewhat cumbersome. Simulink, however, provides for the creation of *subsystem blocks*, which play a role analogous to subprograms in a programming language. A subsystem block is actually a Simulink program represented by a single block. A subsystem block, once created, can be used in other Simulink programs.

You can create a subsystem block in one of two ways, by dragging the Subsystem block from the library to the model window, or by first creating a Simulink model and then "encapsulating" it within a bounding box. We will illustrate the latter method.

We will create a subsystem block for the liquid-level system shown in Figure 3.2. First construct the Simulink model shown in Figure 3.3. The oval blocks are Input and Output Ports. Note that you can use MATLAB variables and expressions when entering the gains in each of the four Gain blocks. Before running the program we will assign values to these variables in the MATLAB Command window. Enter the gains for the four Gain blocks using the expressions shown in the block. You may also use a variable as the Initial condition of the Integrator block. Name this variable h0.

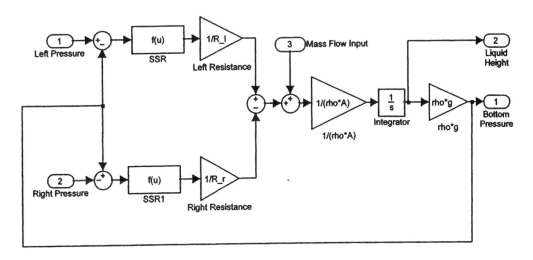

Figure 3.3: Simulink model of the system shown in Figure 3.2.

The SSR blocks are examples of the Fcn block, which is discussed in the textbook. Double-click on the block and enter the MATLAB expression `sgn(u)*sqrt(abs(u))`. Note that the Fcn block requires you to use the variable u. The output of the Fcn block must be a scalar, as is the case here, and you cannot perform matrix operations in the Fcn block, but these are not needed here. (An alternative to the Fcn block is the MATLAB Fcn block, which is discussed in Section 3.11.) Save the model and give it a name, such as Tank.

Now create a "bounding box" surrounding the diagram. Do this by placing the mouse cursor in the upper left, holding the mouse button down, and dragging the expanding box to the lower right to enclose the entire diagram. Then choose **Create Subsystem** from the **Edit** menu. Simulink will then replace the diagram with a single block having as many input and output ports as required and will assign default names. You may resize the block to make the labels readable. You may view or edit the subsystem by double-clicking on it. The result is shown in Figure 3.4.

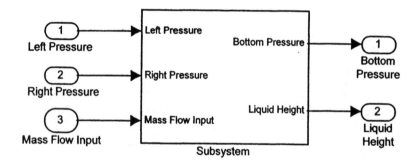

Figure 3.4: Subsystem block based on Figure 3.3.

3.3 Connecting Subsystem Blocks

We now create a simulation of the system shown in Figure 3.5, where the mass inflow rate q is a step function. To do this, create the Simulink model shown in Figure 3.6. The square blocks are Constant blocks from the Sources library. These give constant inputs (which are not the same as step function inputs). The larger rectangular blocks are two subsystem blocks of the type just created. To insert them into the model, first open the Tank subsystem model, select **Copy** from the **Edit** menu, then paste it twice into the new model window. Connect the input and output ports and edit the labels as shown. Then double-click on the Tank 1 subsystem block, set the left-side gain 1/R_1 equal to 0, the right-side gain 1/R_r equal to 1/R_1, and the gain 1/rho*A equal to 1/rho*A_1. Set the Initial condition of the integrator to h10. Note that setting the gain 1/R_1 equal to 0 is equivalent to $R\_l = \infty$, which simulates the absence of an inlet on the left-hand side.

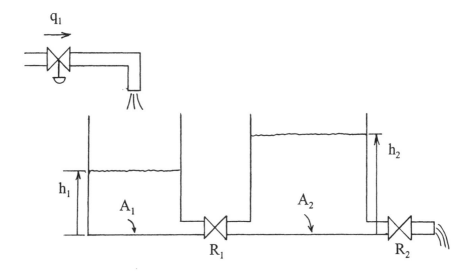

Figure 3.5: A hydraulic system with two storage containers.

Then double-click on the Tank 2 subsystem block, set the left-side gain 1/R_1 equal to 1/R_1, the right-side gain 1/R_r equal to 1/R_2, and the gain 1/rho*A equal to 1/rho*A_2. Set the Initial condition of the integrator to h20. For the Step block, set the Step time to 0, the Initial value to 0, the Final value to the variable q_1, and the Sample time to 0. Save the

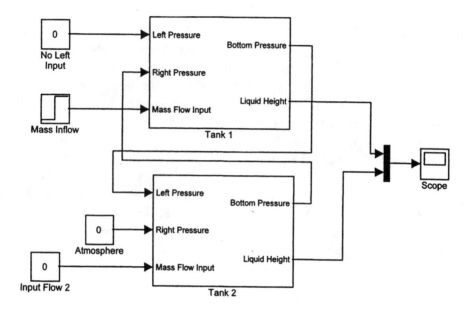

Figure 3.6: Simulink model of the system shown in Figure 3.5.

model using a name other than Tank.

Before running the model, in the Command window assign numerical values to the variables. As an example, you may type the following values for water, in U. S. Customary units, in the Command window.

```
≫A_1=2;A_2=5;rho=1.94;g=32.2;
≫R_1=20;R_2=50;q_1=0.3;h10=1;h20=10;
```

After selecting a simulation Stop time, you may run the simulation. The Scope will display the plots of the heights h_1 and h_2 versus time.

Figures 3.7, 3.8, and 3.9 illustrate some electrical and mechanical systems that are likely candidates for application of subsystem blocks. In Figure 3.7, the basic element for the subsystem block is an RC circuit. In Figure 3.8, the basic element for the subsystem block is a mass connected to two elastic elements. Figure 3.9 is the block diagram of an armature-controlled dc motor, which may be converted into a subsystem block. The inputs for the block would be the voltage from a controller and a load torque, and the

18

output would be the motor speed. Such a block would be useful in simulating systems containing several motors, such as a robot arm.

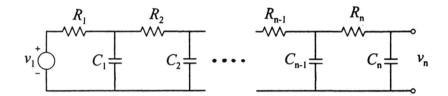

Figure 3.7: An electrical network.

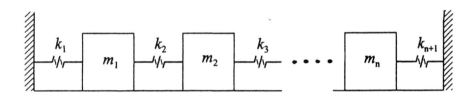

Figure 3.8: A vibrating system.

3.4 Simulation of Systems with Dead Time

Dead time, also called *transport delay*, is a time delay between an action and its effect. It occurs, for example, when a fluid flows through a conduit. If the fluid velocity v is constant and the conduit length is L, it takes a time $T = L/v$ for the fluid to move from one end to the other. The time T is the dead time.

Let $\Theta_1(t)$ denote the incoming fluid temperature and $\Theta_2(t)$ the temperature of the fluid leaving the conduit. If no heat energy is lost, then $\Theta_2(t) = \Theta_1(t - T)$. From the shifting property of the Laplace transform,

$$\Theta_2(s) = e^{-Ts}\Theta_1(s)$$

19

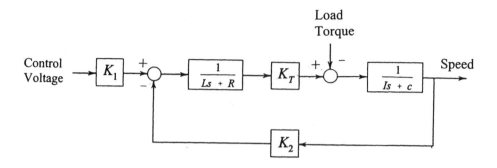

Figure 3.9: Block diagram of an armature-controlled dc motor.

So the transfer function for a dead-time process is e^{-Ts}.

Dead time may be described as a "pure" time delay, in which no response at all occurs for a time T, as opposed to the time lag associated with the time constant of a response, for which $\Theta_2(t) = \left(1 - e^{-t/\tau}\right)\Theta_1(t)$.

Some systems have an unavoidable time delay in the interaction between components. The delay often results from the physical separation of the components and typically occurs as a delay between a change in the actuator signal and its effect on the system being controlled, or as a delay in the measurement of the output. Another, perhaps unexpected, source of dead time is the computation time required for digital control computer to calculate the control algorithm. This can be a significant dead time in systems using inexpensive and slower microprocessors.

The presence of dead time means the system does not have a characteristic equation of finite order. In fact, there are an infinite number of characteristic roots for a system with dead time. This can be seen by noting that the term e^{-Ts} can be expanded in an infinite series as

$$e^{-Ts} = \frac{1}{e^{Ts}} = \frac{1}{1 + Ts + T^2 s^2/2 + \cdots}$$

The fact that there are an infinite number of characteristic roots means that the analysis of dead-time processes is difficult, and often simulation is the only practical way to study such processes.

Systems having dead-time elements are easily simulated in Simulink. The block implementing the dead-time transfer function e^{-Ts} is called the "Transport Delay" block.

Consider the model of the height h of liquid in a tank, such as that shown in Figure 3.1, whose input is a mass flow rate q_i. Suppose that it takes a time T for the change in input flow to reach the tank following a change in the valve opening. Thus, T is a dead time. For specific parameter values, the transfer function has the form

$$\frac{H(s)}{Q_i(s)} = e^{-Ts}\frac{2}{5s+1}$$

Figure 3.10 shows a Simulink model for this system. After placing the Transport Delay block, set the delay to 1.25. Set the Step Time to 0 in the Step Function block. We will now discuss the other blocks in the model.

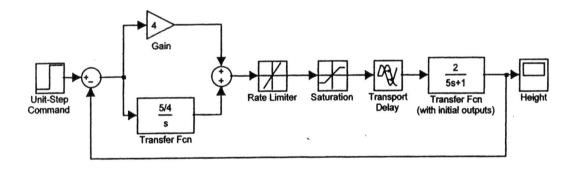

Figure 3.10: Simulink model of a liquid-level control system.

3.5 Specifying Initial Conditions with Transfer Functions

Another block not treated in the textbook is the "Transfer Fcn (with initial outputs)" block, so-called to distinguish it from the Transfer Fcn block. This block enables us to set the initial value of the block output. In our model, this corresponds to the initial liquid height in the tank. This feature thus provides a useful improvement over traditional transfer function analysis, in which initial conditions are assumed to be zero.

The "Transfer Fcn (with initial outputs)" block is equivalent to adding the free response to the block output, with all the block's state variables set to zero except for the output variable. The block also lets you assign an initial value to the block input, but we will not use this feature and so will leave the initial input set to 0 in the Block parameters window. Set the Initial Output to 0.2 to simulate an initial liquid height of 0.2.

3.6 The Saturation and Rate Limiter Blocks

Suppose that the minimum and maximum flow rates available from the input flow valve are 0 and 2. These limits can be simulated with the Saturation block, which implements the function shown in Figure 3.11. After placing the block as shown in the figure, double-click on it and type 2 in its Upper Limit window and 0 in the Lower Limit window.

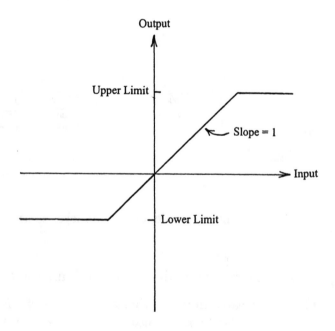

Figure 3.11: The saturation nonlinearity.

In addition to being limited by saturation, some actuators have limits on how fast they can react. This limitation might be due to deliberate restrictions placed on the unit by its manufacturer to avoid damage to the unit. An example is a flow control valve whose rate of opening and closing is controlled by a "rate limiter." Simulink has such a block, and it can be used in series with the Saturation block to model the valve behavior. Place the Rate Limiter block as shown in Figure 3.10. Set the Rising Slew Rate to 1 and the Falling Slew Rate to -1.

3.7 A Control System

The Simulink model shown in Figure 3.10 is for a specific type of control system called a *PI controller*, whose response $f(t)$ to the error signal $e(t)$ is the sum of a term proportional to the error signal and a term proportional to the integral of the error signal. That is,

$$f(t) = K_P e(t) + K_I \int_0^t e(t)\, dt$$

where K_P and K_I are called the proportional and integral gains. Here the error signal $e(t)$ is the difference between the unit-step command representing the desired height and the actual height. In transform notation this expression becomes

$$F(s) = K_P E(s) + \frac{K_I}{s} E(s) = \left(K_P + \frac{K_I}{s} \right) E(s)$$

In Figure 3.10, we used the values $K_P = 4$ and $K_I = 5/4$. These values are computed using the methods of control theory (For a discussion of control systems, see, for example, Ref. 2). The simulation is now ready to be run. Set the Stop Time to 30 and observe the behavior of the liquid height $h(t)$ in the Scope. Does it reach the desired height of 1?

3.8 Simulation of a Vehicle Suspension

Linear or linearized models are useful for predicting the behavior of dynamic systems because powerful analytical techniques are available for such models, especially when the inputs are relatively simple functions such as the impulse, step, ramp, and sine. Often in the design of an engineering system, however, we must eventually deal with nonlinearities in the system and with more complicated inputs such as trapezoidal functions, and this must often be done with simulation. In this section we introduce four additional Simulink elements that enable us to model a wide range of nonlinearities and input functions, namely,

- the Derivative block,

- the Signal Builder block,

- the Look-Up Table block, and

- the MATLAB Fcn block.

As our example, we will use the single-mass suspension model shown in Figure 3.12, where the spring and damper elements have the nonlinear models shown in Figures 3.13 and 3.14. The damper model is unsymmetric and represents a damper whose force during rebound is higher than during jounce (in order to minimize the force transmitted to the passenger compartment when the vehicle strikes a bump). The bump is represented by the trapezoidal function $y(t)$ shown in Figure 3.15. This function corresponds approximately to a vehicle traveling at 30 mph over a road surface elevation 0.2 meter high and 48 meters long.

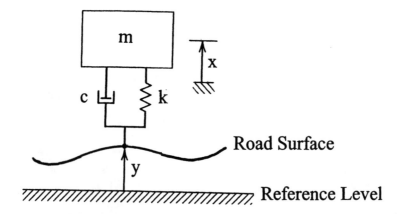

Figure 3.12: Single-mass model of a vehicle suspension.

The system model from Newton's law is

$$m\ddot{x} = f_s(y - x) + f_d(\dot{y} - \dot{x})$$

24

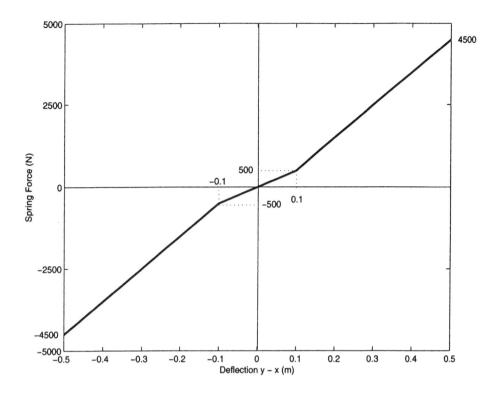

Figure 3.13: Nonlinear spring function.

where $m = 400$ kg, $f_s(y-x)$ is the nonlinear spring function shown in Figure 3.13, and $f_d(\dot{y} - \dot{x})$ is the nonlinear damper function shown in Figure 3.14. The corresponding simulation diagram is shown in Figure 3.16.

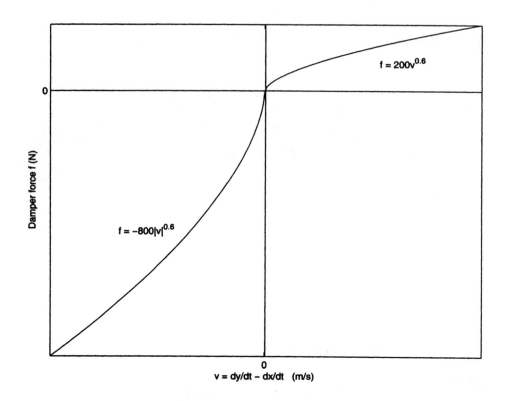

Figure 3.14: Nonlinear damping function.

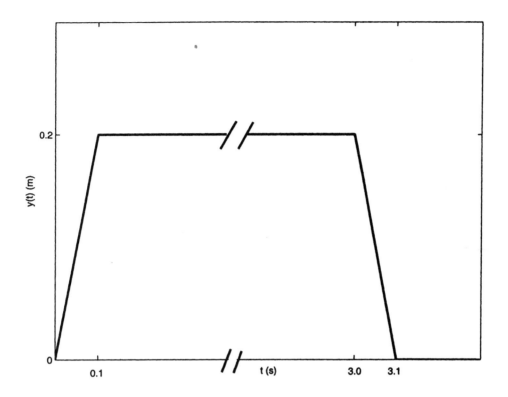

Figure 3.15: Road surface profile.

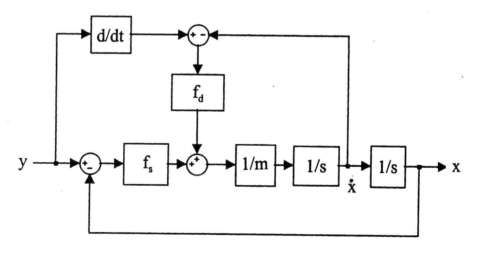

Figure 3.16: Simulation diagram of a vehicle suspension model.

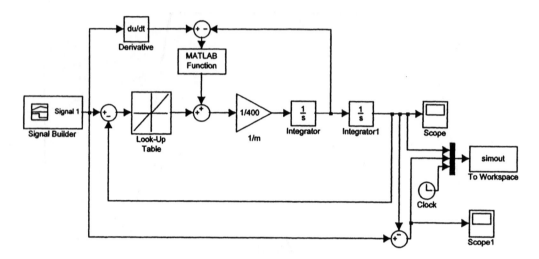

Figure 3.17: Simulink model of a vehicle suspension.

28

3.9 The Derivative and Signal Builder Blocks

The simulation diagram shows that we need to compute $\dot{y}$. Because Simulink uses numerical and not analytical methods, it computes derivatives only approximately, using the Derivative block. We must keep this in mind when using rapidly changing or discontinuous inputs. The Derivative block has no settings, so merely place it in the Simulink diagram as shown in Figure 3.17.

Next, place the Signal Builder block, then double-click on it. A plot window appears that enables you to place points to define the input function. Follow the directions in the window to create the function shown in Figure 3.15.

3.10 The Look-Up Table Block

The spring function f_s is created with the Look-Up Table block. After placing it as shown, double-click on it and enter [-0.5, -0.1, 0, 0.1, 0.5] for the Vector of input values and [-4500, -500, 0, 500, 4500] for the Vector of output values. Use the default settings for the remaining parameters.

Place the two integrators as shown, and make sure the initial values are set to 0. Then place the Gain block and set its gain to 1/400. The To Workspace block and the Clock will enable us to plot $x(t)$ and $y(t) - x(t)$ versus t in the MATLAB Command window.

3.11 The MATLAB Fcn Block

In Section 3.2 we used the Fcn block to implement the signed square-root function. We cannot use this block for the damper function shown in Figure 3.14 because we must write a user-defined function to describe it. This function is as follows.

```
function f = damper(v)
if v <= 0
    f = -800*(abs(v)).^(0.6);
else
    f = 200*v.^(0.6);
end
```

Create and save this function file. After placing the MATLAB Fcn block, double-click on it and enter its name **damper**. Make sure Output dimensions is set to -1 and the Output signal type is set to auto.

The Fcn, MATLAB Fcn, Math Function, and S-Function blocks can be used to implement functions, but each has its advantages and limitations. The Fcn block can contain an expression, but its output must be a scalar, and it cannot call a function file. The MATLAB Fcn block is slower than the Fcn block, but its output can be an array, and it can call a function file. The Math Function block can produce an array output, but it is limited to a single MATLAB function and cannot use an expression or call a file. The S-Function block provides more advanced features, such as the ability to use C language code.

The Simulink model when completed should look like Figure 3.17. You can plot the response $x(t)$ in the Command window as follows:

```
≫x = simout(:,1);
≫t = simout(:,3);
≫plot(t,x),grid,xlabel('t (s)'),ylabel('x (m)')
```

The result is shown in Figure 3.18. The maximum overshoot is seen to be $(0.26 - 0.2) = 0.06$ m, but the maximum *under*shoot is seen to be much greater, -0.168 m.

4. References

1. Palm, William J. III, *Introduction to* MATLAB *6 for Engineers*, McGraw-Hill, New York, 2001, 600 pp.

2. Palm, William J. III, *System Dynamics*, McGraw-Hill, New York, 2004, in press.

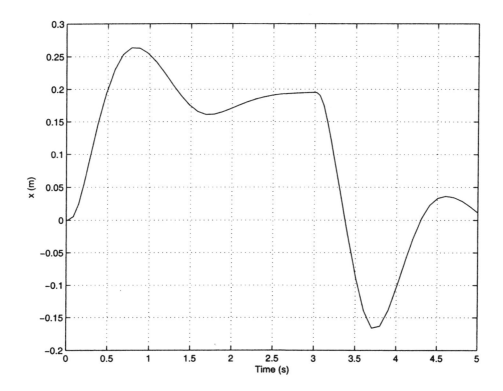

Figure 3.18: Output of the Simulink model shown in Figure 3.17.

NOTES

NOTES

NOTES

NOTES

NOTES

NOTES